# PLOUGHING ENGINES AT WORK

## PLOUGHING / CULTIVATING / DRAINING / DREDGING



**BARRY J FINCH**

**WITH A COMMENTARY ON THE PROCESSES BY THE REV. R C STEBBING**

Published by: Steam Heritage Publishing Ltd, Cranleigh, Surrey, GU6 8HP.

# ACKNOWLEDGEMENTS

I sincerely thank the engine owners, engine drivers, and my friends who have so willingly given their assistance which has enabled me to obtain the photographs and technical details used in this book.

I would also like to acknowledge the help of Paul Proctor A.B.I.P.P. A.R.P.S. for photographic work from an assortment of my old negatives, both film and glass plate.

Also, I would like to thank Brian Gooding for his help and advice in bringing this book into the 21st century.

**Barry J Finch**

**FRONT COVER:** *Windsor* leads *Sandringham* through the Hertfordshire fields. These engines were supplied new to the Ministry of Munitions in September 1919.

**BACK COVER:** The driver makes a regular check of the lubrication, an important item to maintain trouble free running.

ISBN: 978-0-9955432-0-1



Designed and Published by
Steam Heritage Publishing Ltd,
Cranleigh, Surrey, GU6 8HP.

# INTRODUCTION TO THE 2016 EDITION

Over 50 years ago, well known photographer Barry Finch produced a small booklet of more than fifty superb photographs of the twilight days of cultivating by steam, a method that was soon to disappear as the diesel tractor took over, meaning that one man could plough a larger acreage in a day than a team of several men could do with two steam engines.

Although commercial steam dredging continued sporadically into the 1990s, the only places these magnificent engines can be seen working is providing demonstrations at events up and down the country, but the Steam Plough Club, which celebrates its 50th anniversary this year, is keen to keep the skills alive by organising training weekends for today's custodians.

Barry's detailed record of engines working, mainly in East Anglia, gives us a unique and brilliant insight into the last days of regular steam cultivating and the detailed commentary by the late Rev. R C Stebbing in three editions of *Model Engineer* magazine in 1962 is included in this edition to help those with a less intimate knowledge of the subject. We are grateful to the publishers of *Model Engineer* for permission to reproduce this information. (http://www.model-engineer.co.uk)

I am grateful to Kevin Swann for the names and details of some of the crews. His late father was a driver for Taylor Brothers of Wimbish and is seen at work in several of the pictures. A young Kevin even appears in one!

I hope you enjoy this new edition as much as I have enjoyed putting it together with Barry.

**Brian Gooding**
*Editor*

1

Taylor Bros' Fowler BB1 No.15226, *Tiny Tim*, a 16hp class BB1. A typical scene in preparing the engine for work, with the chimney damper hanging off a lamp bracket and the sheets stowed away in the front toolbox. The regular driver, Cyril Richardson, is looking over the motion in preparation to oiling round the many lubrication points. Oil was kept in a quart aluminium teapot for this purpose. Cornelius 'Con' Foley stands awaiting instructions.

# INTRODUCTION TO THE 1962 EDITION

The birth of the traction engine rally has brought together many people interested in steam. More important still, it has saved many engines from the scrap yards, some of which are the sole remaining examples of a particular type or make.

Owners and people who had for a long time been interested in steam have renewed their interest and worked hard to restore and re-decorate their engines to make them suitable to appear at what is now a weekly event during the summer months somewhere in the country.

Many hundreds of thousands of people who had never seen an engine before have become increasingly aware of the sight of a number of engines – anything from half a dozen to nearly a hundred – attending a rally. The sight of an engine on the road under its own steam or even on a low-loader does not create the surprise that it may have done some ten years ago.

Books have been written about traction engines, pictures have been appearing more frequently in the local and national papers, short film reports have been seen on television and a photograph of a beautifully repainted engine has been used as an advertisement by a well known paint manufacturer.

While this interest has been spreading, there has been a small number of engines still employed in the task for which they were made. The majority of this group is, of course, the steam ploughing engine, mostly those built by John Fowler & Co. of Leeds, owned not entirely by enthusiasts but by farmers and contractors, some of whom purchased them new from the makers more than 40 years ago and have worked them more of less every year since.

These engines must rate among the most famous of all steam engines but they have not had the publicity lavished upon them as have the showman's and general purpose agricultural engines, etc. that are more frequently seen at rallies, except perhaps occasionally a picture in a farming magazine. These engines have worked the soil effortlessly but effectively in many cases only 25 miles from the centre of London, hidden from the modern fast moving traffic only by a hedge, the passing motorists not having time to consider the origin of the smoke that tells of a set of steam tackle at work.

The name most widely known in respect of the use of steam cultivation in recent years must be that of the late John Patten, of Little Hadham in Hertfordshire, whose engines worked the neighbouring land until his death and were sold by auction. The sale attracted people from all over England and the sight of a line of eight Fowler ploughing engines in steam will be a sight they will never forget and may well never see again.

The photographs in this booklet unfortunately represent but a few of the engines that have been at work during the last few years but it is hoped they will give some idea of the way a set of steam tackle appears at work to those who have not been fortunate enough to see them. It is also hoped that some pleasure will be derived by those who have known or perhaps been closely connected with steam ploughing during recent or bygone days.

This year, there have been demonstrations of steam cultivation at one or two rallies. Let us hope that the future will see more ploughing engine owners bringing out their great engines to demonstrate to the public or to work on the farms, so that more interest can be encouraged and the art of using steam power in farming is not lost to future generations.

**Barry J Finch**
*October 1961*

# LIST OF PHOTOGRAPHS

# A GUIDE TO METHODS

Too few people these days have had the pleasure of watching steam ploughing tackle carrying out its job in the field. For many reasons, rally demonstrations lack the true atmosphere and may, indeed, suggest that we are looking at some outmoded machinery, whereas steam cultivation remains the most effective method on suitable soil and is only rendered uneconomical by the price of coal.

Now that Barry Finch's fine pictures in 'Ploughing Engines at Work' make it possible for readers to visualise tackle working, it seems opportune to describe in a simple way the machines and their operation so that spectators at rallies may have a better understanding of what they are shown. The numbers in brackets are picture references to the book.

It is best to begin with the implements. In chronological order, the mole draining plough comes first, for John Fowler began using steam power for draining in the early 1850s, though the high wheel type did not appear until 1868. Modified and improved, this is the pattern now in use *(8, 9)*.

The mainframe is carried on two large wheels with a pair of smaller ones in front arranged for steerage through a worm gear and hand wheel at the back. Hinged behind the front wheels is a long beam which can be raised by a chain and hand winch *(9)*. On the beam is a large pulley around which the hauling rope passes, its end made fast to the hind wheel of the engine which is to do the pulling. This double purchase is generally necessary because the work is hard.

Moles are drawn only one way, and the rope from the engine which pulls the plough back is shackled to the end of the beam. The mole itself is a cylinder with a pointed end and is held by a strong coulter projecting at an adjustable depth below the beam *(6)*. When the pull begins, the chain holding up the beam is released and the mole draws itself into the ground *(3)*. As it nears the other engine, speed is slackened and the mole is lifted out of the ground by the winch *(9)*.

On the latest type of drainer, the tail rope does the lifting, instead of manual labour. The plough is then pulled back to the opposite end of the field, the ploughman steering it so that it will be the desired distance from the first drain ready for the next. The engines, of course, move along the headlands while the plough is moving from them.

The mole makes a tunnel in the ground about 4in. diameter at a depth down to about 3ft. These mole drains have been found effective after the incredible lapse of thirty years, though ten or a dozen is more usual. Originally, John Fowler drew wooden pipes strung on a rope after the mole, but it was found that in soil where the mole could be efficiently used, pipes were not needed.

Of course, the important implement in steam ploughing is the plough. It was the Balance Plough designed by David Greig in 1856 which really started John Fowler, with the famous Leeds firm which bears his name, on his successful career. For a reason which will appear, the balance plough is rarely seen today, though a balance sub-soil plough was at Woburn in 1961. Still, its successor can hardly be distinguished from it *(36 to 49)*.

In principle, it consists of two sets of plough bodies coming into work alternately as the plough is drawn to and fro, without turning, between engines on opposite headlands. The breasts, or mouldboards, are right hand at one end and left hand at the other end of the plough, thus turning the furrows all one way.

*Tiny Tim* again, with the mole drainer waiting with George Swann on the steering platform looking for the movement of the flywheel on the far engine. Cyril Richardson is preparing to put water in *Tiny Tim's* boiler via the injector.

# MODIFIED PLOUGH

The plough consists of two main parts. The middle is built of upright and cross members *(37)* with stub axles on either side. On one axle runs the land wheel, bearing on the unbroken ground. One carries the furrow wheel, larger in diamcter, and running in the last turned furrow. These wheels can be swivelled for steerage, a hand wheel being provided at each end of the implement and they can also be raised or lowered in relation to the middle, and with the skid wheels at either end of the frame can regulate the depth of work.

At each end, the frame carries an equal number of plough bodies and is balanced on the middle. The hauling ropes are shackled to the frame below the middle and have the effect of pulling the end of the frame into the ground, bringing the bodies at the trailing end into work, those at the leading end rising into the air.

However, the balance plough has a defect. Unless the depth of working was sufficient, a speed of some 2½ miles an hour was the maximum to prevent the plough jumping out of the ground. At this speed, full use could not be made of the power of the engines and steam could not compete with the horse. Consequently, ploughing contractors were unwilling to work at the 6in. depth customary in England in those days and they ploughed deeper. The plough, of course, turns up the soil and in many instances the soil turned up was far from desirable, and many fields were ruined. This almost killed the use of the plough in England until many years had passed.

Ploughs now in use are fitted with anti-balance gear introduced in the 1880s. The middle is fitted with a pinion on each side engaging with racks formed like shallow inverted Vs in the frame. The ropes are shackled to the middle and can pull it from end to end of the racks, the frame being in balance only when the middle is in the point of the V. The action is hardly easy to describe, while the gear is so hidden that it is difficult to see.

The mole drainer nears Taylor Bros' Fowler BB1 *Old Jumbo*, with Oliver 'Olly' Taylor (Ben Taylor's brother) driving. His hat was very distinctive and could have told many tales of driving all manner of engines, his knowledge was second to none. George Swann on the drainer keeps an eye on the rope on the sheaf wheel in case of problems.

*Old Jumbo* gets some attention from Charlie Taylor, Ben Taylor's brother and partner in Taylor Bros. Olly is in typical stance as the driver, with the mole drainer just entering the ground whilst being drawn away, the large sheaf wheel visible with the double pull of Tiny Tim's rope. George Swann steers the drainer, having tripped the mechanism allowing the drainer blade together with the mole to enter the soil. Many volunteers were recruited for Taylor's steam plough days. Charlie Durrant stands beside the drainer.

5

The mole drainer is pulled away from *Old Jumbo* with the double rope easily visible with the drainer blade at its full depth in the Essex clay soil. Charlie Durrant takes a spell at managing the drainer, while the boss, Ben Taylor, drives one of his engines, together with an old friend, Gerald Dixon, a founding member of the East Anglian Traction Engine Club.

6

The drainer is pulled close to the engine. Charlie Durrant makes an adjustment to the trip mechanism which holds the beam containing the long blade and attached mole head. Ben Taylor driving, appears about to take out the plough dog clutch which takes the drive to the rope drum under the boiler. Note the big wooden block hanging on the rear of the tender horn where the balance plough would normally be coupled when travelling on the road.

**7** Another shot of the mole drainer being pulled away from *Old Jumbo*.

A superb view of the mole drainer at full working depth, the double rope and sheaf wheel again visible. George Swann stands with one boot on the end of the drainer beam, the other on the small platform fitted for steering.

Here both George Swann and Charlie Durrant are working the handles of the chain lifting mechanism whilst walking backwards as it nears the end of the pull. Some mole drainers had a self-lifting arrangement, no such luck here.

10

Fowler AA7 No.14726, aptly named *Peace*, for, along with her sister engine *Victory*, No.14727, they were both delivered on the day the Armistice was declared in November 1918. The driver, Arthur Martin, had the nick name of 'Pickles'. He had a clubbed right hand, although he could handle any engine as good as many others. He is in typical steam ploughman's pose with one boot on the edge of the tender, keeping a watchful eye on the cultivator as it nears the end of its pull close to the headland of the field. This class of Fowler ploughing engines were some of the largest to be used in this country.

11

*Victory* nears the end of its pull with the cultivator, the driver, Jim Letch, readying himself as the implement nears the engine. The trailing rope of the other engine can be seen on the far side of the cultivator. This type of work was often dusty as can be seen from the dust cloud behind the cultivator steersman, Harold Jackson. This was made worse when the field was cultivated both length and breadth ways, with little or no comfort save for a folded sack on the steersman's seat come toolbox.

*Victory* is moved forward in readiness for the next pull, the turning of the cultivator being one of the most dangerous moments as the frame of the cultivator is lifted from the soil by the pull of the opposite engine's *(Peace)* rope. We see the driver closely watch as the implement turns, in readiness to whistle to stop his mate on the other engine. Both drivers would be watching for steam to go up from the whistle as it was quick actions that were needed in these cases. With the sound of the whistle having to travel, sight proved the best method. Again the crew are Jim Letch and Harold Jackson.

13 Driver Arthur Martin on *Peace* at Elms Farm has a short break whilst the other engine takes the cultivator away across the field. It would appear he has just added more coal to the fire in preparation for the next pull. His left hand is on the injector steam tap probably topping the boiler up with water.

Jim Letch on *Victory* steadies his drum speed as the cultivator, with Harold Jackson steering, nears the engine.

Nearing the end of another pull, the cultivator steersman, Harold Jackson, is just about to turn the steering in readiness for being pulled back by the opposite engine.

# PULLING THE PLOUGH

**The operation of the plough at a headland needs explaining.**

**The sequence of events.**

The implement is drawn as close as possible to the engine, the leading end, of course, in the air *(40, 41)*. The middle is nearer the leading end. When the direction of pull is reversed, the middle rolls along the rack to the point of the vee and the plough is then in balance *(37)* and can be swung into the next row of furrows *(42)*. As the tail of the plough drops, bringing the bodies at that end into the ground, the middle rolls forward and transfers an additional weight to the tail end.

Ploughs have been built for many purposes in many lands, and a great variety of breasts have been devised, but for this country, the bevel frame type with breasts for shallow or seaming work, or with semi-digging type for medium depth, has generally been employed. Ten-furrow ploughs have been built for shallow ploughing, but five or six furrows are mostly used for medium work.

Before the First World War, the most popular implement in England was the turning cultivator *(11 to 31)*, the forerunner of which appeared in 1868. Two sizes can be found, 9/11 tine, or 11/13. Tines are long prongs which cut into the ground and break it up, but do not turn the subsoil to the top.

Two large wheels running on cranks at each end of an axle carry the rear end of a frame fitted with tines. A smaller wheel, swivelling for steerage under the control of the rider, is in front *(17)*. Above this wheel is the turning lever which has three arms. That at the back has a chain attached to a segment fixed on the axle, the hauling ropes being shackled to the other arms which point forward.

At the end of a bout, the rope taking up the pull from behind the cultivator causes the turning lever to swing round: the cranked axle is given a partial turn by the chain, lifting the frame and thus the tines out of the ground *(12, 16)*, while another chain from the turning lever locks the front wheel round.

The lifting segment has teeth engaging with a pawl on the upright catch lever and when the turn is complete, the rider, having steered into line, pulls back this lever and allows the tines to drop into work. Many cultivators have a dashpot fitted to lessen the shock of this drop.

It is customary to cultivate two ways, say north and south, with crossings running east to west. The second time, the pulverising of the soil is increased by the speed of the implement.

In all probability, most people regard the ploughing engine (PE) as an overgrown traction engine with a coil of rope under the boiler, and imagine it to be built to give various power outputs.

A study of the history of design shows that the PE has borrowed little from traction engines. Rather the reverse, for it has developed independently.

It was realised early on that the number of men needed to work tackle in the field – for example, two drivers, at least one man to ride the plough, with a man or boy with horses for the watercart – was the same for three acres an hour as for two. Consequently, the largest set possible was the most economical, the limiting factor being the size of implement conveniently handled. Generally, this involved engines too big for average farm work, hence the PE is really a locomotive windlass.

Fowler BB1 No.15436, *Princess Mary*, was new to the Executors of William G Fairhead, Peldon, Essex in late July 1920. It is seen here, in the ownership of Mr P Yorke, cultivating land in Sussex. The 'A' shaped frame carrying both ropes is visible, together with the steering wheel on sharp lock aiding the turning of the cultivator, this being a Fowler 11/13 tine, here with 11 tines fitted. Bill Druce is on the cultivator.

Compared with a traction engine, the PE is not too well adapted for either hauling loads or belt work; the power outputs quoted as being vastly greater than those of traction engines are misleading. They are "indicated h.p." figures "while ploughing"; that is, while the engine is running at speed for bouts of, say, four minutes, during which time no feed water is added nor fire touched. With the traction engine, power output figures must be in respect of continuous working.

In the design of a PE, the following factors must be taken into account: size and type of implement normally used. (This affects ploughing gear ratio, for the cultivator should run faster than the plough), size of fields (length of rope needed) and nature of soil. These factors have a bearing on cylinder and boiler sizes.

Just prior to the 1914-18 war, at least twenty varieties of engine were being built at Leeds; in fact, five types are illustrated in this book!

The ancestor of the AA was built in the 1880s, and was among the first compound engines. Since there was no home trade at that period, the AA was designed with a big boiler and ploughing speed on the low side, the plough being the implement most likely to be employed.

With a revival in the home market came the BB in 1913, having a higher rope speed to suit the cultivator; two cylinder sizes were adopted, one the same as the AA, giving the same power with lesser pull on the rope, the other having smaller bores, frequently found in conjunction with double speed ploughing gear. The BB1, in 1919, was altered in detail and had a larger firebox.

During the war period,  the AA appeared with some variations. First there was a larger cylinder previously used for the ZA, then came a more up-to-date engine, the AA7, and later an AA externally as the AA7 but with the old original AA cylinder dimensions.

Returning to the book, 14383/4, *Prince* and *Princess*, are BB: 15226/7 are BB1: 14726/7 are listed as AA (with ZA cylinders), 15364/5 are AA7, while 15563 is again AA.

A driver's eye view of a Fowler cultivator. The short rope attached to the steering drum frame helps when turning at the end of each pull. The quality of work is improved by the speed the cultivator is drawn through the ground. Six mph can be safely reached.

The late John Patten's 16hp Fowler class BB No.14383, *Prince*, cultivating land in Hertfordshire in October 1959. This engine, together with her sister *Princess*, was to only see another season of commercial work before being sold into preservation. Both driver and set foreman Walter 'Wally' Hammond (standing on engine running board) keep their eyes out for any possible problems.

The cultivator has just entered the ground after turning. The coal trailer is filling the bunker of *Princess*, this being a regular job when these engines were working hard together, and bringing water with ever more frequency. On long days, this could be up to four or five times for both engines.

The luxury of working from a hard road was not the norm. A selection of short chains hang from the tender in readiness.

**21** Smoke rises into the autumn sky as *Princess's* driver steadily pulls the cultivator. The neatly folded sheet for covering the engine overnight sits on the front box where the 'Clips', sometimes referred to as 'Spuds', which are fitted to the rear wheels when working on soft and muddy ground.

*Prince* stands gently simmering, showing all working day needs of the steam ploughmen, such as chains and wooden blocks, which could be required at a moment's notice.

*Prince* again. The large rope drum can carry up to 1,000 yards of ¾in. steel rope. The water lifter hose is neatly coiled from the foot board above the drum; the chimney damper typically hangs from the lamp bracket on the smokebox.

Taylor Bros' *Old Jumbo* with their Fowler 11/13 tine cultivator nearing the end of another pull. George Swann is about to turn the steering on the implement, Charlie Taylor, partner in Taylor Bros, stands in front of the engine. Oliver Taylor is driving. He had a lifetime of driving engines in many areas of England & Wales. Lastly Maurice Gowlett, Oliver's son-in-law, whose wife Peggy could drive engines and rollers as good as most men.

25

Taylor Bros' Fowler *Tiny Tim* finishes another pull with the cultivator on fields near their yard at Wimbish. Cyril Richardson, in typical stance, is driving with Charlie Durrant watching from the steersman's box; George Swann is with the cultivator.

26

Pulling out the steel rope is helped by Olly holding the clutch lever in on *Old Jumbo*. Charlie Taylor is nearest the engine, while George Swann pulls on the rope head preparing to attach it to the 'Y' frame of the cultivator. In the background we can partially see Broad Oaks Manor, an Elizabethan house with its own history.

*Tiny Tim* nears the end of another pull with George Swann taking charge of the implement and Cyril Richardson in typical pose. Ben Taylor stands to the left of the group, together with Charlie Brown, a founding member of the East Anglian Traction Engine Club, Reg Swann (George's brother), lastly Jim Cliff, a local farmer and engine owner from nearby Ashdon.

28

A rarely seen shot with Ben Taylor and two of his brothers – Olly driving *Old Jumbo*, Charlie standing on the engine's running board, with George Swann with his brother Reg on the cultivator. *Tiny Tim* is at the far end of the field some 500 yards away.

29

No.15227, *Old Jumbo*. The high speed at which the flywheel turns when the engine is pulling is obvious. Olly is in typical pose whilst driving. In working days, where the men were paid acreage money, it was the norm to run these engines at speed (old habits die hard).

John Patten's Fowler AA7 18nhp No.15354, *Windsor*, simmers in the sun during a quiet moment. The cultivator shows the less than luxurious seat covered in an old sack as padding on the toolbox top. The boiler tube brush and rod protrude from its resting place beside the boiler. Near Sawbridgeworth, Hertfordshire, June 1957.

31

Another view of *Windsor* and the cultivator.

*Windsor* during a quiet moment, the driver attending to oiling the many lubricating points of the engine. Note the large oil cans forward of the steering wheel where the contents would be warmed in readiness. The bundle of straw is in preparation for lighting up, while the large tree trunk attached to a chain hanging from the balance plough (horn) coupling would be trailed behind the rear wheel as a block to stop the engine running back when working on hills. Unusually, these engines are fitted with a brake to the rear wheels, although the handle appears to be missing.

18hp class AA7 compound No.15353, *Sandringham*, negotiates the headland of the field with *Windsor's* rope attached in preparation for cultivating the ground a second time, this being diagonally to previously. The clinker shovel and rake are seen hanging from the brake handle attached to the tender.

34

A typical scene where a set of steam tackle, having completed one job, prepares to move on to another. With the anti-balance plough already coupled to *Windsor*. *Sandringham* is readied for coupling to the crew's living van, where the team lived whilst away for up to five or six days, and cultivator. Note the motorbikes which are probably the only means of transport for the men. Hertfordshire, October 1959.

35 *Windsor* taking the lead from *Sandringham*, with the tarpaulins neatly folded on the front toolbox.

# PLOUGHING WITH ROPE AND DRUM

To distinguish between the classes, the easiest way is to look at the slide bars; the BB and BB1 have only one to a cylinder *(23)*. The BB has large false covers on the back of the cylinder *(55)*; the BB1 has an extension to the low pressure piston rod and the false covers are smaller *(1)*. (A ploughing engine has the back of the cylinder towards the chimney, and the boiler front plate corresponds to a locomotive backhead!)

The AAs have two slide bars to a cylinder *(53)* and the shape of the cylinder block differs from that of the BB, which is in proportion higher (see *57*: the engines in the foreground are BBs and the remainder AAs).

There are other differences, not very easy to discern. Nos.14726 and 14727 differ from the others in their road gearing *(10)*. The general rod gear clutch on a ploughing engine employs a sliding slow-speed pinion on the crankshaft with dogs on its outer side. The fast-speed pinion runs loose in constant mesh with the second-motion double wheel; to engage the fast speed, the dogs on the slow-speed pinion pick up corresponding dogs in the fast pinion when the slow-speed is moved outwards. Originally both pinions ran on the crankshaft. The fast one was called, from its shape, the "bell" pinion *(10)*. It is held axially by collars and could, if desired, be taken off during ploughing, saving wear and tear and a bit of noise. On the AA7 and BB, the slow pinion was combined with a sleeve on which the fast pinion was loose, being held axially by a kind of cage *(53)*.

The ploughing gear is naturally of first importance. It consists of a drum rotating on a stud fixed below the boiler barrel. As well as rope, the drum has a ring of gear through which it is driven: a pair of bevels transmitting the motion of the crankshaft by means of the upright shaft which has a clutch and pinion at its lower end *(52)*. The clutch is a dog, to drive one way only, and reversal of the engine automatically disengages it; in fact, at the end of a bout, the custom is to slow down and reverse at the last moment.

As the rope must coil on the drum on the side away from the firebox, the clutch, and therefore the engines, must be built right-hand and left-hand. When the rope comes out on the left, the engine is called the left-hand engine and when it is ploughing, runs in forward gear: the right-hand engine runs in reverse.

Should it be necessary to pay out some rope, the driver can help by holding the clutch lever in the engaged position *(26)*.

Projecting from below the drum is the coiling lever to guide the rope so that it coils properly. If the coils are not right, shock can be felt on the engine and damage be done to the rope. The coiling gear is designed for a definite number of coils: if it is for eleven coils deep, the rope will he guided from top to bottom of the drum while the drum makes eleven turns. It follows that the correct size of rope to make eleven even coils on the drum must always be used. The depth of drum varies and the length of rope ranges from 600 to 1,000 yards for AA and BB engines. The pull on the rope may be from four to five tons.

BB engines may be found with double-speed ploughing gear in the shape of two bevels on the crank and upright shafts operated by a lever beside the ordinary clutch lever.

The ploughing engine is easy to handle in spite of its size, though the absence of compensating gear can affect the steering. It is advisable to avoid locking sharply, as either front or hind wheels must skid; in the field, angle-iron segments ("toe nails") are bolted to the front wheel rim to stop the wheels from sliding through loose ground.

36

No.15354, *Windsor*, couples up to a plough before moving on to the next land to be worked.

Setting tackle down to work is perhaps not as simple as may be thought. Fields may have awkward corners, and the way to cover them has to be planned. The appearance of the finished work is spoilt if the ropes have to drag over it. It is easier to avoid priming and to keep the top of the firebox covered if the engines face uphill. A steam ploughman is quoted in the RASE's Review in 1867: "Begin ploughing at the bottom of the field, turning the furrows downward: begin cultivating, short way first, at the top."

A typical operation in the field would be roughly as follows. Engine A hitches on the rope from engine B and moves across to the far headland *(33)* and gets set for its first pull. A's rope is shackled to B's rope and B pulls back until both ropes can be shackled to the implement which has been standing close to B. A now pulls for the first bout.

As the plough nears the engine, the driver eases the speed. With one hand on the regulator and the other on the reversing lever *(40)*, at the last moment he puts the reverse over and closes the regulator. The engine, if the timing is right, will just turn over the other way, disengaging the clutch. The road gear is engaged and the engine moves forward to be in line for its next pull. Engine B now pulls gently while the ploughman with help *(42)* swings the plough ready for its return.

With a cultivator, the rider lets go of his steering wheel until the return pull has lifted the tines up and brought the cultivator nearly round, when he resumes steering *(11, 12)* and pulls the catch lever, allowing the tines to drop into work.

Driver A turns on his injector (safety valves may be lifting, but water will be low) and in all probability he will need to replenish the fire, for during a bout he dare not interfere with either for fear of losing steam. He may keep the engine running gently to draw the fire up. The injector is very useful for this method of working, and it is very rare to find a ploughing engine built after about 1912 with a pump.

These notes have been written just to give a general idea of ploughing engines and their work as far as they are illustrated in this book: they are not to be taken as an exhaustive treatise on the subject!

37

The lines of a Fowler Anti-balance plough car be appreciated here. These ploughs were in common use throughout most areas of England during the heyday's of steam ploughing (1910-1950s).

Wimbish, Essex, September 1961. Taylor Bros' *Old Jumbo's* silhouette and smoke appear through the mist, with their six-furrow plough. Cyril Richardson steers while Maurice Gowlett on the tail makes adjustments to the plough depth as required.

39 The plough midway through another bout, the art of the ploughman (Cyril Richardson) clearly visible by the straight furrow being maintained.

40

Cyril on *Tiny Tim* pulls the plough close up to the engine with Olly Taylor's watchful eye looking over his shoulder. George Swann on the plough turns out of the furrow in readiness for moving the plough into the next pull.

41

*Old Jumbo* completes another pull with Frank 'Daddy' Saville driving, Cyril Richardson and Maurice Gowlett on the plough.

42

Many hands make light work! The plough is being slowly pulled out of balance by *Old Jumbo* at the other end of the field, the plough crew and bystanders pulling it down. George Swann on *Tiny Tim* is about to move forward so that its rope can clear the plough breasts and is hooked on the plough tail. Olly walks away from the scene in his forever-seen wellington boots.

43

Cyril on No.15226, *Tiny Tim*, with Olly Taylor watches as the plough starts its outward run. George Swann takes charge on the plough assisted by Maurice Gowlett. Wimbish, Essex, September 1961.

44

The plough is about at full depth with the frame near the straw stubble. George Swann's clear concentration is on maintaining his straight furrow while Maurice Gowlett adjusts the furrow wheel. Charlie Taylor catches a lift on the tail end.

Frank 'Daddy' Saville looks out from driving with a young Kevin Swann taking notes whilst sitting on the steering box. Cyril Richardson and Charlie Durrant are on the plough as Con Foley prepares to bring in the water cart.

46

George Swann with *Tiny Tim* in a quiet moment, working close up to the headland, the chimney damper hanging in its normal place on the lamp bracket.

An unusual view of a Fowler six-furrow plough during its journey across the field with George Swann in charge on the plough assisted by Maurice Gowlett. The rope can be seen under the plough with the trailing rope of the other engine behind.

The plough approaches *Tiny Tim* at the end of a pull. The smoke from the engine breaks the sky on a misty day.

A Fowler plough comes out of the mist. Under such weather conditions, the drivers usually work purely by whistle signals. These old drivers would know what to expect through years of working commercially carrying out the work. Some say it was telepathy!

A close up view of the drum of *Prince* carrying 1,000 yards of rope. The clutch and rope guide pulleys can easily be seen.

Following the heyday of steam ploughing, a number of companies offered their services for dredging lakes and marshland where draglines and diggers could not reach. Here we see a bucket scoop at work, the engines being stationed on either side of the area to be cleared.

Another view of the scoop. This type of scoop is capable of moving a large amount of mud at each pull.

Gears and motion of No.15364, *Windsor*. With years of hard work showing, *Windsor's* driver, Walter 'Wally' Hammond, still kept her brasswork polished. Note the dents in the cylinder cover, likely to have been caused when the plough has not been stopped in time.

54

During a quieter moment, Walter 'Wally' Hammond checks the lubrication of *Windsor's* motion, important to maintain trouble free running.

55

John Patten's steam plough gangs still take pride in their engines, ensuring they were clean on the day where they were auctioned off. L-R: unknown, Pat Lawrence, unknown, Walter 'Wally' Hammond, unknown, unknown. In the background are *Prince* and *Princess*.

Sale Day, 28th June 1960, at John Patten's farm, Little Hadham. Four pairs of engines were in steam and sold with other steam tackle. Although four sets of engines where sold – *Prince & Princess*, *Lion & Tiger*, *Sandringham & Windsor*, *Haigh & French* – the first three sets were saved to enter preservation.

Proud drivers clean their engines before the crowds arrive for the sale at Little Hadham, Hertfordshire. Working from memory together with Steam Plough Club records, the derelict engine standing at right angles to the others would be either Fowler AA6 No.13727, *Beatrice*, Reg: NK 982, or No.13728, Reg: NK 987. New on 02/11/1917 to George Fox, Great Munden, Hertfordshire. Later to John Patten, both derelict in 1959. From the sale they went to J W Hardwick & Sons, West Ewell, Surrey and were scrapped.

A fine AA7 18hp Fowler, restored to original condition. No.15563, *Daisy*, supplied new in 1920 to Lord Rayleigh, Terling, Essex. Now owned by Mr P Yorke and renamed *Wayfarer*. A fine example of Fowler's supreme AA7 engine.

Another look at the front cover picture, this time uncluttered by the title: *Windsor* leads *Sandringham* through the Hertfordshire fields. These engines were supplied new to the Ministry of Munitions in September 1919.

# PLOUGHING ENGINES AT WORK

PLOUGHING / CULTIVATING / DRAINING / DREDGING

# BARRY J FINCH

Now married for 65 years and the father of two sons, Barry Finch was born in 1930 in Dunton Green, Kent, and was educated at Sevenoaks Preparatory School and the Judd School in Tonbridge.

In 1949, being called up for National Service, he trained and served as a photographer in the Royal Air Force. After demobilisation, he joined the Esso Petroleum Film & Photographic Department, and later spent 20 years as a staff photographer with the Courtauld's Film Unit. He served for several years on the committe of the National Traction Engine Club, as it was then.

His interest in steam engines, like many other people, started as a young boy watching the engines threshing on the local farms. In 1955, he purchased the Wantage engine No.1389 from its original owner, saving it from the hands of the scrap merchants Hardwick's of Ewell, who had also bid for it.

Photography, as well as being his work, was his hobby and this is when his picture collection really began. When the steam fairs and rallies came on the scene, he was fortunate to have many of his pictures published, particularly in *The Model Engineer*, and it was their sales representative who suggested that Barry publish a book of pictures. This was how *Ploughing Engines at Work* was born.

The other titles he produced were:

- *A Rally of Traction Engines*
- *Traction Engines in Review*
- *Traction Engines in Colour*